Rafaela Neris Gaspareto
Marcelo C.M.T. Filho

Egeria densa as a nutrient substrate for ornamental sunflowers

Rafaela Neris Gaspareto
Marcelo C.M.T. Filho

Egeria densa as a nutrient substrate for ornamental sunflowers

The use of aquatic macrophytes in the nutrition of ornamental sunflowers with or without mineral fertiliser

ScienciaScripts

Imprint

Cover image: www.ingimage.com

This book is a translation from the original published under ISBN 978-3-330-76171-1.

Publisher:
Sciencia Scripts
is a trademark of
Dodo Books Indian Ocean Ltd. and OmniScriptum S.R.L publishing group

120 High Road, East Finchley, London, N2 9ED, United Kingdom
Str. Armeneasca 28/1, office 1, Chisinau MD-2012, Republic of Moldova, Europe
Managing Directors: Ieva Konstantinova, Victoria Ursu
info@omniscriptum.com

Printed at: see last page
ISBN: 978-620-8-37506-5

I dedicate this book to my family, friends and all those who have supported me and believed in my potential.

ACKNOWLEDGEMENTS

To God, for giving me the strength to always keep going and for everything I've achieved and the victories I've won, including this one and the ones I'm yet to achieve.

In order for this work to be carried out and completed, it was necessary to involve people to whom I would like to express my sincere thanks:

To my family, for helping and supporting me, including my mum Ivone, my sister Karine and all the other family members;

To Prof Dr Marcelo Carvalho Minhoto Teixeira Filho, for guiding and mediating this important step in my career;

To my friends, Mariane and Benedito, as well as everyone who helped make this work a reality.

"To acquire knowledge, you have to study; but to acquire wisdom, you have to observe." (Marilyn vos Savant)

SUMMARY

After analysing the nutrients and toxic elements contained in the aquatic macrophyte Egeria densa, it was found that there is potential for its use as an organic fertiliser and that the dried and partially decomposed macrophyte showed physical characteristics similar to those of commercial substrates. The aim was therefore to evaluate the use of the macrophyte Egeria densa as a nutrient substrate, associated or not with conventional or controlled release mineral fertiliser, in the development and nutrition of ornamental dwarf sunflower (Helianthus annuus L.). The statistical design was entirely randomised with four replications, arranged in a 5 x 3 factorial scheme: five doses of the macrophyte Egeria densa (0, 25, 50, 75 and 100% of the volume of the substrate in the pot, and the rest made up in proportional terms with a commercial substrate with a pH of 6.0-6.5) and three fertilisations (conventional fertiliser, slow release fertiliser (Osmocote®) and no mineral fertiliser) carried out when the sunflower was sown. The plot consisted of one sunflower plant per 4-litre plastic pot, with a spacing of 0.45 m between rows. Increasing the doses of the macrophyte Egeria densa in the substrate had a positive effect on the levels of N, P, K (leaves+flower stem), Mg, S, Mn, leaf chlorophyll index (LCI), plant height, flower stem and flower disc diameters, vegetative dry matter and the dry matter of the ornamental sunflower chapter, regardless of mineral fertilisation. However, the highest doses of this macrophyte reduced the Cu and Zn content (leaves+flower stem) and root dry matter. The use of the macrophyte Egeria densa as a nutrient substrate in a proportion of between 25 and 40% was feasible for the production of ornamental sunflowers with standards suitable for marketing both in pots and for cutting, provided that it was associated with conventional mineral fertilisation or with osmocote.

Key words: Helianthus annuus L., Organic fertiliser, Improved efficiency fertiliser, Ornamental plant.

SUMMARY

CHAPTER 1

INTRODUCTION

The cultivation of flowers in Brazil has been growing in recent years and is a promising segment of intensive horticulture in the national agribusiness field, thus increasing its influence on the country's economy through the commercialisation of flowers for the foreign market (CURTI, 2010).

Studies combining the cultivation of ornamental sunflower (Helianthus annuus L.) with aspects of plant nutrition and soil fertility are scarce. It is therefore necessary to carry out research in this area in order to determine the parameters of fertiliser sources that are best suited to this crop. One alternative has been the use of controlled release fertilisers, particularly Osmocote®, which has been widely used in ornamental sunflower cultivation. One of the explanations for the increased use of this fertiliser is that water penetrates the granules, dissolving the nutrients which are gradually released from micro-cracks in the coating polymer into the substrate where the plant's root system is located, depending on its temperature and humidity.

For ornamental dwarf sunflower, as with other species, the nutrient most in demand is nitrogen (N), and the quantity, form and cost of the fertiliser used must be taken into account. In floriculture, the cost/benefit ratio of the dose of N is important for the viability of the enterprise, with the relevant factors being the quality of the fertiliser, its solubility and reaction in the solution (BRAGA, 2009). In view of the change in consumer behaviour and their concern for sustainability, the use of aquatic macrophytes (Egeria densa) as an organic fertiliser makes the product stand out and contributes to environmental preservation.

According to Correa, Velini and Arruda (2003), the average nutrient levels found in the E. densa species, especially N, P, K and Ca, indicate that this plant can be used as a nutritional source for different crops, in addition to its use for animal consumption.

In research carried out by Sampaio and Oliveira (2005), it was found that using the mass of Egeria densa as an organic fertiliser for maize was viable and recommended, as it showed higher agricultural productivity compared to the other treatments analysed that used cattle manure. Becaleto et al. (2014), after analysing the nutrients and toxic elements contained in the macrophyte Egeria densa collected from the Paraná River, near the Ilha Solteira Hydroelectric Power Station, also found great potential for its use in plant nutrition as a biostimulator and fertiliser, elucidating its possible use in agriculture.

Aquatic macrophytes have proliferated a lot in the reservoir of the Ilha Solteira Hydroelectric

Power Station, probably due to the eutrophication of the waters of the Paraná River, to the point where they have become unwelcome on the city's beaches. For control, there are reports of herbicides being applied.

In view of the above, there is a need for further research to see if the use of the aquatic macrophyte Egeria densa as a nutrient substrate or to enrich commercial substrates is viable and it is also interesting to compare its use without or with conventional or controlled release mineral fertiliser.

The aim of this study was therefore to evaluate the use of Egeria densa macrophyte as a nutrient substrate, with or without conventional or controlled-release mineral fertiliser, in the development and nutrition of ornamental sunflowers.

CHAPTER 2

LITERATURE REVIEW

2.1. FLOWER PRODUCTION AND MARKETING IN BRAZIL

According to França and Maia (2008), the main market for flowers in Brazil is the domestic market, although this has a low per capita consumption of around US$ 4.70 per inhabitant, while Switzerland consumes US$ 170 per inhabitant. Brazilian consumption therefore has great potential for expansion, both for cut flowers and potted flowers.

The flower market in Brazil has seen great growth (Table 1). According to data from the Brazilian Institute of Floriculture (IBRAFLOR 2015) in 2014, the cultivated area was around 14,992 hectares, with more than 350 different species, including more than 3,000 varieties, and per capita consumption was R$26.68 per inhabitant. As such, the flower market is an important cog in the Brazilian economy, responsible for 215,818 direct jobs: 78,485 (36.37%) related to production, 8,410 (3.9%) related to distribution, 120,574 (55.87%) in retail and 8,349 (3.8%) in other functions, mostly as support (IBRAFLOR, 2015).

Table 1 - Flower **market** in Brazilian regions (year 2014)

Region	Number of producers	Area (ha)	Job creation
South East	4018	8561	132962
South	2232	2714	36709
North-East	1138	2027	26579
North	437	861	7985
Centre West	423	829	11583
Total	8248	14992	215818

Source: Adapted from IBRAFLOR (2015).

As the flowers and ornamental plants market has gained prominence in the national territory, its participation in the Gross Domestic Product (GDP) has been verified, and in 2014 this figure was R$4.51 billion (Table 2), referring to the Flowers and Ornamental Plants production chain in Brazil. Sectoral GDP was calculated using the sum of sales of the final products in the production chain. It should also be noted that this type of product can be found in all sectors, so it's easy for consumers to buy, which can lead to an increase in consumption and consequently influence the sector's production chain and the values generated in the country's economy.

Table 2-Estimated gross domestic product (GDP) of Brazil's flower and ornamental plant production chain in 2014

Product	Internal market (IM)	External market (ME)	Total (MI+ME)
	R$	R$	R$
Floristry	984.330.709	-	984.330.709
Decoration	2.340.728.679	-	2.340.728.679
Landscaping	649.395.492	-	649.395.492
Self-service	385.161.923	-	385.161.923
Wholesale to the end consumer	120.700.722	-	120.700.722
Producer to end consumer	60.265.288	-	60.265.288
Export		55.958.381	55.958.381
Import		(-) 83.004.272	(-) 83.004.272
Total	4.540.582.815	(-) 27.045.891	4.513.536.924

Source: IBRAFLOR (2015).

In Brazil, 90% of the formal marketing of national floriculture products takes place in the state of São Paulo. This trade takes place in large wholesale centres such as the Veiling Holambra Cooperative, the CEASA/Campinas Permanent Flower and Ornamental Plant Market, CEAGESP in São Paulo and Cooperflora/Floranet (JUNQUEIRA; PEETZ, 2008).

According to Neves (2003), the commercialisation of potted plants is one of the most value-adding options in the horticulture sector, and species, varieties or hybrids are often selected for the market in order to create differentials in their production.

According to research carried out by Silva (2012), the main goods sold in floristry are potted flowers, with 92.2%, followed by cut flowers and potted and cut foliage. With this data, it can be seen that ornamental sunflowers, with their exuberant flowers and rapid flowering, fit into the main forms of commercialisation, and can be sold in pots or as cut flowers.

When it comes to buying and selling flowers, the quality of the product is the main preponderant factor in the choice, and some characteristics of the flower are involved, such as colour, the number of green leaves present, the number of flower buds, among others. All these attributes are evaluated by consumers and traders according to each species. For 49.5% of the wholesalers interviewed by Silva (2012), the quality of the flower is paramount, followed by price and colour.

Neves et al. (2009) pointed out that, with floral art techniques constantly changing due to market trends, the consumer's refined taste is diverse and is not limited to the most traditional and publicised flowers. As a result, flower growers must present new species as alternative crops in order to remain at the forefront of knowledge and supply. As a cut flower, the sunflower offers many shades of bloom, harmonising well with other species and foliage, hence its great demand in the flower arranging market.

To produce flowers and ornamental plants, there is a high demand for inputs, which increases the value of the final product. Companies supplying inputs to the flower and ornamental plant production chain had a turnover of approximately R$856 million in 2014, as can be seen in Table 3. You can also see the turnover of each type of input company.

Table 3-Estimated turnover for each group of input companies in the Flowers and Ornamental Plants Production Chain in 2014

Input companies	**Turnover in 2014 (R$)**
Seedlings, seeds and bulbs	248.551.116
Substrates	171.109.632
Fertilisers	82.623.708
Defences	30.713.865
Biological control	3.326.540
Pruning and harvesting tools	3.277.093
Personal protective equipment (PPE)	3.574.900
Vases	152.310.805
Packaging	98.829.070
Utilities and companies linked to water supply	4.774.083
Heating fuels	24.334.833
Electricity utilities	32.529.775
Total input turnover	855.955.420

Source: FERREIRA and BELO (2015).

Analysing the data above (Table 3), it can be seen that the main expense in the production of flowers and ornamental plants is the use of seedlings, seeds and bulbs, followed by substrates and fertilisers. It is therefore essential to reduce these amounts if the production chain is to be more profitable, as in most cases these inputs are not reused for more than a year and their perishability is their main characteristic.

In this context, costs can be reduced by replacing the common substrate with another product that can fulfil the same function and provide the nutrients needed for the development and growth of flowers, including ornamental sunflowers. In this way, the substrate could be partially replaced by a species of aquatic macrophyte, for example, and with the possibility of also reducing the cost of fertilisers, which is the third largest source of income for input companies (Table 3).

1.1. ORNAMENTAL SUNFLOWER (Helianthus annuus L.)

The first indication of commercial sunflower cultivation in Brazil was registered in 1902 in the state of São Paulo, when the state's Department of Agriculture distributed the first seeds to farmers. After that, in the 1930s, the sunflower was used for various purposes, such as forage and silage production, mellifera, seed production for edible oil extraction and poultry feed (UNGARO, 1986).

The sunflower is an annual plant, botanically classified in the Asteraceae family. It originated in North America, more precisely in south-west Mexico, where it grows naturally. The species was introduced to Europe in the 14th century as a cultivated plant and reintroduced to America from Europe in the 19th century (SALUNKHE ; DESAI, 1986).

The species Helianthus annuus L. has a chapter-type inflorescence, made up of generally sessile flowers, which are formed at the apex of the stem, elongated discoid, forming a receptacle where the flowers are inserted. The receptacle has hairy and rough bracts and the diameter of the chapters can vary depending on the species, climate and soil (LENTZ et al. 2001).

The ornamental sunflower comes from the hybridisation of the graniferous sunflower. Its use has increased with the diversification of the colour of its flowers (PELEGRINI, 1985).

Through the technique of directed artificial hybridisation, hybrid varieties of sunflower have been launched with shades of rust, burgundy, pink, light pink, yellow with orange and lemon yellow mixtures, with a dark and/or light disc (OLIVEIRA;

CASTIGLIONI, 2003). It is important to note that, in order to sell ornamental sunflowers, the flowers on the disc must be sterile, thus making pollen production impossible and having characteristics suitable for the consumer market.

The CNPSoja Ornamental Sunflower Programme, which began in 1996, has created nine shades of flower colour, providing economical alternatives for use in gardening and making flower arrangements, and the plants are adapted to Brazilian climatic conditions (RIBEIRO et al., 2007).

Several hybrid cultivars of ornamental sunflower have been developed for cut flowers and, with some modification to the cultivation methods, they can be used as pot flowers. Neves et al. (2009) mentioned some of these cultivars, such as Sunrich Gold, Sunrich Lemon, Sunrich Orange, Moonbright, Sunbeam, Sunbright and Sunbright Supreme. There are also cultivars on the market specially developed for pot cultivation, such as Big Smile, Elf, Ted Bear, Sundance Kid, Sunspot and Pacino Gold.

Brazil has great climatic variations throughout its territory, making it suitable for growing different plant species, including the ornamental sunflower, which is widely accepted in the flower market due to its colouring and peculiar shape. This species is suitable for both pot and cut plant commercialisation, catering for different types of consumers.

1.1.1. CULTIVATION

Flower production and flower and stem size are characteristics defined by genetic potential and can be influenced by the mineral nutrition and cultural treatments used on the crop (HIGAKI; IMAMURA; PAULL, 1992).

The best sowing time for ornamental sunflower is when it is possible to meet the edaphoclimatic requirements such as climate, soil, water availability and temperature characteristic of each region for the plants in the development phases, resulting in a reduction in the risk of diseases appearing, especially after flowering, thus ensuring good productivity, as well as making it possible to plan for cutting and utilisation (CURTI et al., 2012). The length of the vegetative cycle can vary depending on the cultivar, sowing date and environmental conditions. Other characteristics, such as the sunflower's height, chapter size and stem diameter, vary depending on the genotype and the plant's soil and climate conditions.

For maximum sunflower development and growth, a minimum air temperature of 10°C at night and a maximum temperature of 25°C during the day are recommended, with the optimum temperature for development being 18°C. In addition, the plant in question is highly adaptable to different types of soil; however, it grows best in soils with a moderately acidic to neutral pH, greater than 5.2 (determined in CaCl2) and with good drainage (SIMÃO, 2004).

The sunflower plant's water needs vary from less than 200 mm to more than 900 mm per cycle, with a consumption range of between 500 and 700 mm, well distributed throughout the cycle. A lack or excess of water is detrimental to plant development. The most critical phases for water deficit are between 10 and 15 days before the start of flowering and grain formation (DOORENBOS; KASSAM, 1979).

When producing ornamental sunflower in pots for commercialisation, one of the challenges is to guarantee adequate plant growth and high aerial biomass production, especially of the chapter, with a limited volume of roots, restricted to a small volume of substrate. Lemaire (1995) observed that the physical, chemical and biological properties of substrates play an important role in ensuring the mechanical maintenance of the root system in the solid phase, the supply of water and nutrients in the liquid phase, as well as the supply of oxygen and the transport of carbon dioxide between the roots and the outside air in the gaseous phase.

The correct use of fertiliser when planting ornamental sunflowers, with the use of N, P and K, green manure and top dressing, will allow the plants to have larger receptacle sizes (SFREDO; CAMPOS; SARRUGE, 1984). It should also be noted that the action of nutrients in the critical phase of flower differentiation determines the potential number of flowers, which is why it is important to provide the plant with an adequate supply of nutrients through the substrate and fertilisers.

In the crop, the period in which the highest rate of nutrient absorption and fastest growth occurs is between the formation of the flower bud and the complete expansion of the inflorescence (EVANGELISTA; LIMA, 2008).

Curti et al. (2012) pointed out that, for marketing ornamental sunflowers, the appropriate harvest point is when the chapters have 50% of the ray's ligulate flowers open. These stems should be cut as long as possible and the leaves should be removed from the lower half, as well as any that are damaged. The flowers should be placed in disinfected containers containing clean water and preservative.

The ideal sunflower for cut flowers should essentially produce chapter sizes smaller than 10 cm in diameter, because if they are too large, when used for ornamentation (in floral arrangements and/or bouquets), they can deform the flower stalks due to their weight (SIMÃO, 2004).

For marketing, the flower stalks are grouped into bundles of five, selected according to the height of the stem and the diameter of the chapter(s). The bundles must be kept in a cool, dry environment until they are sold. The Hellianthus genus has great post-harvest durability, so the flower stalks end up keeping for longer periods in ornamentals (SIMÃO, 2004).

In order to assess the quality of ornamental sunflowers, both the morphological aspects of the plant must be taken into account, such as the flower structure, shape, length and diameter of the chapter, number of flowers and buds, as well as their colour, and the absence of chemical residues, pests, diseases and apparent deformations (CURTI et al., 2012).

In order to be accepted on the market, sunflowers must have a quality standard. In general, the standard established by Veiling Holambra is used for the flower grown in pots (Table 4) and for cutting (Table 5).

Table 4-Criteria for classifying the quality of potted ornamental sunflowers

	Standard vase A1	**A2 standard vase**
Minimum plant height	20 cm	15 cm
Maximum plant height	40 cm	45 cm

Height tolerance	5 cm	10 cm
Rods	Firm, straight and with good support	
Disease damage (Botrytis, rust)	0	Light intensity without compromising the beauty of the product
Pest damage (caterpillar, aphid, whitefly and mite)	0	Light intensity without compromising the beauty of the product
Mechanical damage	1 to 2 leaves per pot	More than 2 leaves per pot without jeopardising plant formation
Yellow leaves	0	2 leaves per pot starting to yellow
Phytotoxicity burn	0	Light intensity in the leaves without compromising the beauty of the product
Chemical waste	0	Light intensity in the leaves without compromising the beauty of the product

Source: IBRAFLOR (2015).

Table 5-Criteria for classifying the quality of ornamental sunflower for cutting

Class	**Stem length (cm)**	**Thickness (cm)**	**Open flower diameter(cm)**
50	50	Min. 0.8	Min. 6,0
60	60		
70	70		
80	80	Min. 1,1	Min. 7,5
90	90		

Source: IBRAFLOR (2015).

2.3. CONTROLLED RELEASE FERTILISERS

In order to produce Helianthus annuus L., it is necessary to fertilise in such a way as to provide the plant with the right amount of nutrients at the right time. In other words, the nutrients must be available when the plant species needs them most for its development, so that the plants can complete their cycle with the best performance. This can be done using conventional fertilisers or controlled release fertilisers, which will be discussed next.

The low fertility conditions of Brazilian soils, combined with the high nutrient requirements of crops, end up resulting in large losses or inadequate use of fertilisers in the soil/plant system. One alternative for reducing these losses would be to spread fertiliser applications over the crop cycle, but these operations increase production costs and the need for labour. Another alternative to this problem is the use of controlled release fertilisers, such as Osmocote®, where research is increasingly trying to reconcile this technology with the requirements of the various crops grown in Brazil.

Controlled release fertilisers are those in which the nutrients are released to the plants gradually. This process is due to a layer of organic resin, which regulates the release of these nutrients. Water vapour penetrates the fertiliser granules and dissolves the nutrients, which are gradually released into the substrate depending on its temperature and humidity (ZAMUNÉR FILHO, 2009).

Higher temperatures favour faster nutrient release and consequently slow down the release period of the Osmocote granules® . The release of nutrients from Osmocote® , according to the manufacturer, is not influenced by pH, water quality, substrate type, external salt concentration or microbiological activity (SCOTTS UK PGB, 2002).

The fertilisers described are very practical and responsive, with different formulations and periods for making nutrients available to the plants, which can be 3 months, 6 months and so on. A continuous supply during the plant's growth period provides lower leaching losses and a higher concentration of nitrogen in the tissues, with greater growth compared to the use of highly soluble fertilisers (CARVALHO, 2001). However, there is still little research to prove these benefits in terms of plant nutrition, production and quality, especially those studying ornamental sunflowers.

2.4. Egeria densa

2.4.1. IMPORTANCE

Macrophyte species such as Egeria densa, which are considered weeds, find the same survival conditions in nature as non-weed species. Therefore, what differentiates them from others and determines their success are the biological and physiological strategies they have acquired that allow them to exploit the environment in an opportunistic and competitive way (PIERINI; THOMAZ, 2004). Thus, when observing the behaviour of this species, it is considered to be a plant that can cause imbalance in the aquatic environment, causing damage to the environment.

The large volume of aquatic plants, especially in times of high rainfall, favours the proliferation of these organisms. In view of this, it was noted that it was necessary to reduce the maximum operating power of the Jupiá - SP reservoir plant by up to 60 per cent so that the grids would not be clogged due to the presence of Egeria densa, which would consequently cause hydraulic pressure on the grids and could increase the damage (PRINCIPE; KURATANI; MELONI, 1997).

Marcondes et al. (2000), in their research, observed that the Companhia Energética do Estado de São Paulo (CESP) has been suffering losses at the Jupiá Hydroelectric Power Station since 1994 due to the high infestation of aquatic plants, causing a 9.1% reduction in electricity generation in 1999. It is therefore clear that the presence of macrophytes in power station reservoirs causes losses in energy production.

Another aspect of the high incidence of macrophytes in aquatic environments is ecological imbalance and, consequently, eutrophication, which is a process characterised by the natural and gradual fertilisation of surface waters, resulting in increased biological productivity due to the high concentration of these organisms. In addition, the presence of algae and/or macrophytes leads to difficulties in river transport, fishing and also makes it impossible to use water bodies for leisure.

During their life cycle, submerged macrophytes assimilate large quantities of nutrients. These are released in small quantities into the aquatic environment during the period of active growth, and after senescence and decomposition the remaining nutrients are released in greater quantities (WETZEL, 1993). Based on this characteristic, the species Egeria densa, which is a submerged macrophyte, can be considered as a potential source of nutrients with a low C/N ratio, since this plant is generally discarded on hydroelectric dam sites.

2.4.2. DESCRIPTION, DEVELOPMENT AND COMPOSITION OF Egeria densa

Firstly, in order to use a species as a substrate, it is necessary to know its morphology, physiology and composition, so that it can be used appropriately and fulfil its function in the study. This paper will present the characteristics of the macrophyte Egeria densa, so that we can understand its use as a source of nutrients for plant production.

The species studied is classified as an aquatic macrophyte, which is a terrestrial plant adapted to the aquatic environment, submerged and perennial, belonging to the Angiosperm group and the Hydrocharitaceae family, being a monocot. E. densa is a species native to South America, introduced all over the world and widely found in rivers and lakes, with the ability to form dense populations. For its growth, it needs clear, clean waters, with a mild temperature (20-24 °C) and rich sediment at the bottom, reaching two to three metres in length (NASCIMENTO et al., 2001).

It is dioecious, with white, small, pedicellate flowers, around 2.0 cm in diameter. Rodrigues, Dettke and Montanher (2007) reported that the plant has uninerved leaves, the stomata are absent, the margins are serrated, verticulate or opposite towards the base of the stem and some epidermal cells on the margin have a denticular shape; they are covered with a thin cuticle and reduced to just two cell layers, and have a group of fibres. They have thick walls, no lignin deposition and consist mainly of cellulose.

The stem of E. densa is straight, with dichotomous branching and made up of a unistratified epidermis, whose cells are elongated axially and covered by a thin cuticle. The cortical cells contain a large number of starch grains, especially in the cells close to the central cylinder, which are used as a reserve substance by aquatic macrophytes. Adventitious roots are smooth and unbranched, and starch grains can also be found inside the cortical cells, but to a lesser extent than in the stem (RODRIGUES; DETTKE; MONTANHER, 2007).

2.4.3. USES FOR Egeria densa

The species E. densa, present in large quantities in bodies of water, can be harmful to the environment in which it is found. However, there has been an increase in the number of research studies and publications employing other uses for these plants, such as animal feed and plant nutrition.

In research carried out by Módenes et al. (2011), it was found that the macrophyte under study had great potential for removing the reactive dye blue 5G, which is present in industrial textile effluents, as there was a significant reduction in the amount of dye present in the analysed solution, using the algae in an ecological manner.

The organic waste from aquatic macrophytes can be used to produce compost and is a good raw material. Since there is no other good use for them when they are removed from the water body, this process can take between 55 and 65 days, according to research carried out by Silva et al. (2009).

In experimental trials carried out by Silva et al. (2013) to assess the effectiveness of using aquatic macrophytes as a source of organic matter for the germination of mastic tree seeds, it was found that this type of plant provided suitable conditions for seed germination and the development of seedlings of this tree species.

Sampaio, Oliveira and Nascimento (2007) in their research with maize, using organic fertiliser with cattle manure and Egeria densa, concluded that the mass of the macrophyte released greater quantities of nutrients than the manure, from the first days of incorporation into the soil. The apparent recoveries by the plants of the amounts of N, P and K applied were also higher than those of the manure and reached 39, 103 and 76% respectively, thus recommending its use as a fertiliser.

Valeri and Corradini (2000) pointed out that, due to the different types of materials used to make up the substrate, it is important to determine its physical characteristics so that it can be used in a way that provides an appropriate ratio between the volumes of its constituents in the tube or in any other place where it is stored, i.e. volume of air, volume of water and volume of solids.

Growing plants in alternative substrates has been increasingly used in our country. The substrates must be low cost, available close to the region of consumption, have sufficient nutrient content, good cation exchange capacity, allow aeration and moisture retention, and favour the physiological activity of the roots (OLIVEIRA; HERNANDEZ; JUNIOR ASSIS, 2008).

Therefore, the use of Egeria densa as a nutritious substrate in the production of flowers and ornamental plants becomes an alternative to conventional substrates, as it has the desirable characteristics for the development of plants such as ornamental sunflowers. The use of this alternative substrate will be presented in this work as a proposal to reduce input costs (substrates and fertilisers) in a more sustainable way.

CHAPTER 3

MATERIAL AND METHODS

3.1. STUDY AREA

The municipality of Ilha Solteira is located in the north-west of the state of São Paulo, 650 km from the capital, with geographical coordinates of 20°38'44" South Latitude and 51°06'35" West Longitude. According to the Koppen system, the climate is classified as AW (tropical), the average annual rainfall is 1,300 mm and the average annual temperature is 28°C.

Regional altitudes range from 280 to 380 metres. Today, the municipality's main economic activities are, in hierarchical order: electricity, livestock farming and agriculture.

The experiment was carried out in a greenhouse with an anti-aphid screen and 50% shading, at Campus II (Agronomy) belonging to the Faculty of Engineering - UNESP, located in the municipality of Ilha Solteira - SP.

3.2. RESEARCH METHODOLOGY AND CONDUCT

The aquatic macrophyte (Egeria densa) was collected manually from the bed of the Paraná River, upstream of the Ilha Solteira Hydroelectric Power Station. After collection, it was dried in full sun for 7 days on a tarpaulin with a layer that was not too thick and then ground in a mechanical grinder with a 5 mm sieve. The nutrient content of the E. densa used in this research were: 26.0 g kg^{-1} of N, 1.9 g kg^{-1} of P, 26.5 g kg^{-1} of K, 11.1 g kg^{-1} of Ca, 5.9 g kg^{-1} of Mg, 4.6 g kg^{-1} of S, 23 mg kg^{-1} of B, 16 mg kg^{-1} of Cu, 1872 mg kg^{-1} of Fe, 713 mg kg^{-1} of Mn and 30 mg kg^{-1} of Zn, determined according to Malavolta, Vitti and Oliveira (1997), at UNESP - Câmpus de Ilha Solteira. The levels of toxic elements found in the seaweed were: 0.81 g kg^{-1} of Na, 0.0 mg kg^{-1} of Pb, 4862.0 mg kg^{-1} of Al, 1.2 mg kg^{-1} of Cd and 14.5 mg kg^{-1} of Cr, which were determined at ESALQ - USP.

After analysing the nutrients and toxic elements contained in the aquatic macrophyte Egeria densa, it was found that there is potential for its use as an organic fertiliser and that this dried and partially decomposed macrophyte has physical characteristics similar to those of commercial substrates.

The statistical design was entirely randomised with four replicates, arranged in a 5 x 3 factorial scheme: five doses of the macrophyte Egeria densa (0, 25, 50, 75 and 100% of the volume of the substrate in the pot, and the rest made up proportionally with a commercial substrate with a pH of 6.0-6.5) and three different fertilisations (conventional fertiliser, slow release fertiliser (Osmocote®) and no mineral fertiliser), carried out when the sunflower was sown. The plot consisted of one sunflower plant per 4-litre plastic pot, with a spacing of 0.45 m between rows.

The pots were filled with commercial substrate and/or seaweed substrate on 28 February 2015. The substrate has the following characteristics: pH 6.0-6.5 composed of pine bark, coconut fibre, vermiculite and rice husk.

In the treatments in which the conventional substrate was mixed with the aquatic macrophyte substrate, the latter was incorporated and homogenised. After this procedure, the containers were left in the greenhouse and watered a few times before sowing the ornamental sunflower to prevent the aquatic macrophyte from fermenting, which could damage the seeds and/or seedlings.

Sowing took place on 9 March 2015. Initially, two seeds were sown per pot. After ten days, it was noticed that none of the plants had germinated in some of the pots. Therefore, on 21 March, the treatments in which the two seeds had germinated were thinned out and these were carefully transplanted into the containers in which the seeds had not germinated.

Mineral fertilisation was carried out at the time of sowing and was based on the sunflower's requirements and the recommendation to use Osmocote® Plus 15-09-12, which, according to the manufacturer, has a controlled release with longevity for local climatic conditions, varying from 1 to 4 months after application and an application recommendation of 4.7 g L^{-1} of substrate.

As Osmocote® has been widely used to grow ornamental plants in conventional substrates, this was the reference treatment. Therefore, the doses of nutrients from conventional mineral fertilisers (0.42 g of P2O5 (Triple Superphosphate), 0.71 g of N (Ammonium Sulphate), 0.56 g of K2O (Potassium Chloride), 0.06 g of Mg (Kieserite) were equivalent to those supplied by Osmocote® . There was no need to apply S because ammonium sulphate and kieserite contain this nutrient in their composition.

In order to meet the water needs and ensure the good development of the ornamental dwarf sunflower, manual irrigation was carried out twice a day, once in the early morning and once in the late afternoon, as required. It was found that the treatments containing the greatest amount of Egeria densa retained the most water and did not need a second watering during the day.

After handling the crop in the greenhouse (Figure 1), there was a significant reduction in the volume of substrate in the treatments with 100% E. densa, which may have been due to the density of the substrate and the amount of water retained. Fungi were found in some of the treatments that contained more algae, which could be explained by the fact that when the aquatic macrophyte was removed from the aquatic environment and dried, at some point, either during this process or during storage, contamination by spores could have occurred and, by providing suitable conditions of temperature and humidity, the fungi developed. When the fruiting bodies appeared, they were removed manually and did not cause any damage to the ornamental sunflower seedlings.

Figure 1 - Crop **development** in the greenhouse.

Source: author

Although it was carried out in a greenhouse, two applications of Decis 25® were made on 8 and 25 April 2015 to control bedbugs.

Cultivation was similar for all the treatments, but the treatments: 100% seaweed with conventional fertiliser, 75 and 100% seaweed with slow release fertiliser had dead seedlings. There is no conclusion as to what could have caused this. One explanation could be that the seaweed fermented and thus prevented the plant from growing, causing it to die. However, this is only an inconclusive hypothesis.

The first assessments were carried out on 18 May 2015, when most of the plants were already in full bloom (Figure 2). The second assessment, when the inflorescence had not yet fully opened, was carried out 7 days later (25 May 2015), the crop cycle being 70-84 days. The assessments carried out in these

The dates were as follows: leaf chlorophyll index (LCI), which is calculated by the content of light transmitted by the leaf, using two or three wavelengths and different absorbances. A portable digital chlorophyll meter was used for this purpose, and the readings were taken on the last newly expanded leaf in the morning; plant height, with the aid of a graduated ruler; total number of dry leaves and green leaves per plant; disc diameter, considering the inner part of the chapter without the bracts; flower stem diameter, with the aid of a digital caliper, measured from 1 cm above ground level.

Figure 2- Ornamental sunflower with fully open flower.

Source: author

After collecting all this data, the plants were cut close to the substrate. The roots, vegetative aerial part and inflorescence were then separated, the roots washed to remove the accumulated substrate and all placed in an oven at 105° C for 72 hours to dry and then weighed to obtain the respective dry matter.

The dry matter of the flower stalks and leaves, the shoot, the roots and the total of the sunflower plants were determined. These samples were then ground in a Wiley-type mill to determine the macro and micronutrient content of the vegetative aerial part and the chapter, according to the methodology described by Malavolta, Vitti and Oliveira (1997). These analyses were carried out in the Plant Nutrition laboratory of the Faculty of Engineering of UNESP - Ilha Solteira Campus.

3.3. STATISTICAL ANALYSIS

The results obtained were analysed using analysis of variance and Tukey's test at 5% probability to compare the means of the mineral fertilisers, and regression equations were adjusted for the effect of the doses of aquatic macrophyte. The statistical analysis programme used was SISVAR.

CHAPTER 4

RESULTS AND DISCUSSION

4.1. QUALITATIVE AND BIOMETRIC ANALYSES

With regard to the qualitative analyses, the following items were classified: leaf chlorophyll index (LCI), plant height, flower disc diameter, flower stem diameter and number of green leaves and dry leaves (Table 6), relating the doses of Egeria densa and the mineral fertilisers applied or not.

Analysing the data, it can be seen that for the leaf chlorophyll index (LCI) and the number of dry leaves, the mineral fertilisations and the control did not differ significantly. For the diameters of the floral disc and the floral stem, the values of the treatment without fertilisation were lower than those fertilised with osmocote or conventional fertiliser. With regard to plant height, it was noted that the controlled release fertiliser did not show any significant difference between conventional fertilisation and the treatment without fertilisation. With regard to the number of green leaves, Osmocote showed a greater number than conventional fertiliser and no fertiliser.

Table 6- Leaf chlorophyll index (LCI), plant height, floral disc and floral stem diameters and quantities of green leaves and dry leaves of ornamental mini sunflower as a function of doses of the macrophyte Egeria densa and mineral fertiliser.

	ICF	**Plant height** (cm)	**Disc diameter** (cm)	**Loral stem diameter (cm)**	**Num. Dried leaves**	**Num. Green leaves**
Doses of E. densa **(% of substrate volume)**						
0	28,188	61,388	7,775	0,900	5,917	20,083
25	31,275	66,916	6,758	0,900	4,667	23,583
50	34,725	66,083	8,358	0,950	4,750	22,250
75	32,325	57,000	8,750	0,869	4,625	22,250
100	39,125	33,500	4,500	0,475	3,500	14,500
Fertilisation						
Without	31,968 a	54,583 b	6,245 b	0,728 b	4,500 a	18,550 b
Conventional	31,050 a	66,625 a	8,150 a	0,947 a	4,936 a	20,875 b
Osmocote	34,100 a	63,750 ab	8,562 a	1,012 a	5,500 a	26,833 a
MSD (5%)	3,409	10,315	1,199	0,150	1,834	3,054
Overall average	32,195	60,889	7,494	0,872	4,896	21,396
CV (%)	11,960	19,130	18,060	19,370	42,300	16,110

Means followed by the same letter in the column do not differ according to Tukey's test at 5% probability.

When analysing the leaf chlorophyll index, there were higher values in the 50 and 100% doses of the Egeria densa macrophyte as a substrate (Figure 3). This high index is related to the amount of nitrogen present in the plant, as this nutrient is a constituent of the chlorophyll molecule, indicating that the macrophyte tested supplied N to the ornamental mini sunflower, and consequently influenced photosynthesis. Analysing the statistical breakdown, it was observed that at the 50% macrophyte dose there was a significant difference (Figure 4), with the use of the controlled release fertiliser having a higher leaf chlorophyll index.

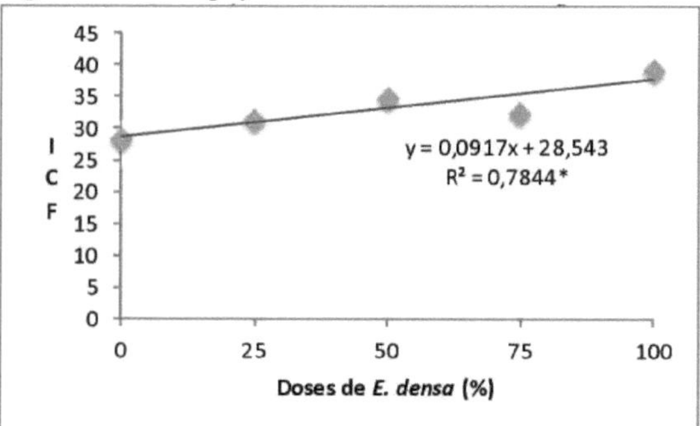

Figure 3: Leaf chlorophyll index (LCI) of ornamental mini sunflower as a function of doses of the macrophyte Egeria densa.

Source: author herself.

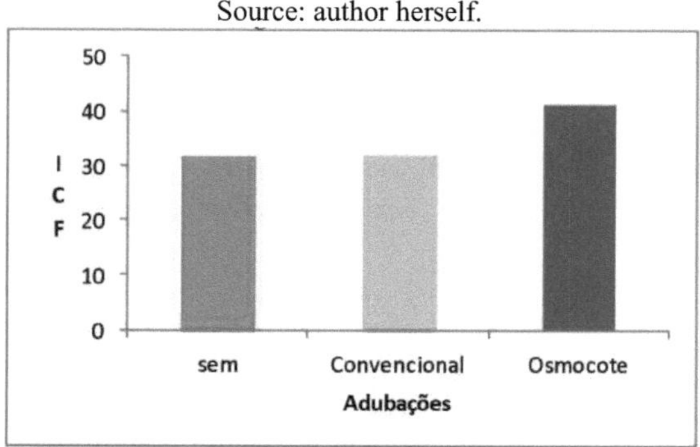

Figure 4. ICF in ornamental mini sunflower from the 50% dose of the macrophyte Egeria densa.

Source: author herself.

With regard to plant height, there was an increase up to the estimated dose of 32.5% of E. densa present in the substrate (Figure 5). This can be explained by the fact that the volume of the macrophyte decreased due to irrigation, causing it to decompose and the volume of the substrate to decrease compared to the other treatments. However, even with the reduced size of the plants with the 100 per cent dose, they can be classified as potted plants and the others as cut flowers.

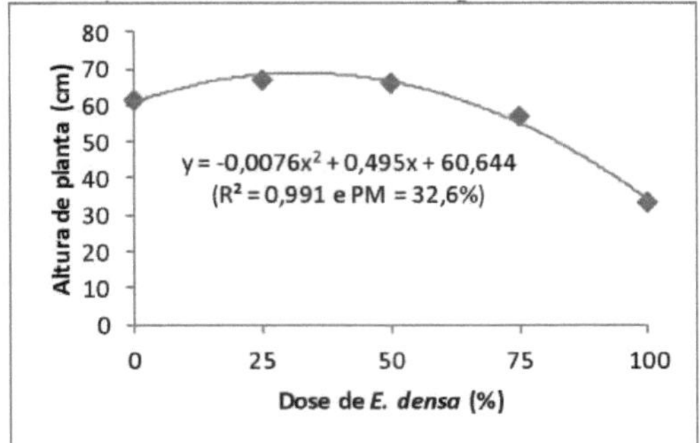

Figure 5 - Plant height of ornamental mini sunflower as a function of doses of the macrophyte Egeria densa.

Source: author herself.

For the diameter of the flower disc in which the petals were not counted (Figure 6), the optimum dose estimated was 39.8% of the substrate composed of E. densa. It is also worth pointing out that even the smallest flower bud size can be considered suitable for commercialisation. The presence of fertiliser, whether conventional or with osmocote, was found to promote larger flower sizes (Table 6).

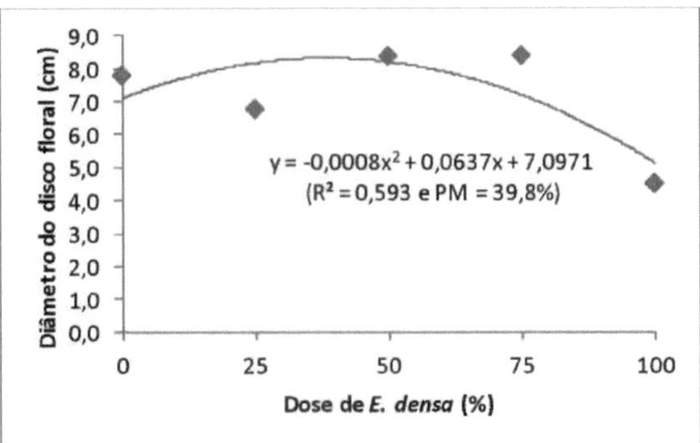

Figure 6. Flower disc diameter of ornamental mini sunflower as a function of doses of the macrophyte Egeria densa.

Source: author herself.

The diameter of the flower stem of the ornamental mini sunflower increased up to the estimated dose of 35 per cent of the substrate with E. densa (Figure 7). This assessment is important because the floral stem contains the conductive vessels that carry nutrients and water to the plants. It was observed that the diameter of the floral stem was similar to the diameter of the floral disc, i.e. there is proportionality between the two pieces of data; the larger the inflorescence, the larger the diameter of the floral stem, as there is a greater need for nutrients.

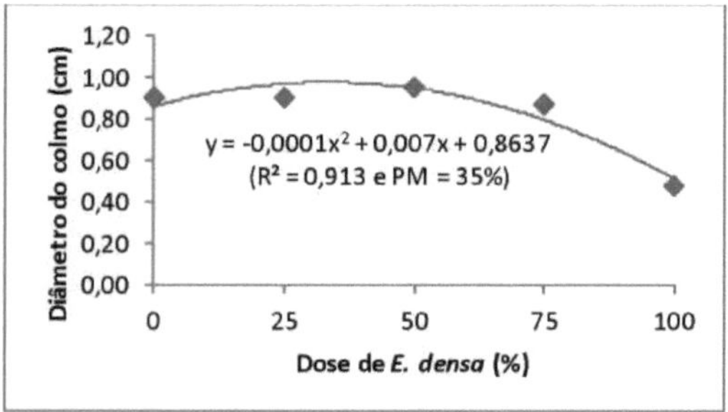

Figure 7- Diameter of the flower stem of ornamental mini sunflower as a function of doses of the macrophyte Egeria densa.

Source: author herself.

When buying an ornamental plant, consumers take into account the visual aspect, which includes the quantity of dry leaves, i.e. the greater the quantity, the lower the commercial value of the plant, and it may even be discarded because it doesn't meet quality standards. With this in mind, it can be seen that the amount of dry leaves was inversely proportional to the dosage of Egeria densa (Figure 8), where the greater the amount of the macrophyte as a substrate, the lower the amount of senescent leaves. In addition, the treatment with the largest volume of macrophyte was the one that provided characteristics for commercialisation in pots, with a low quantity of senescent leaves, which did not interfere with the quality standard.

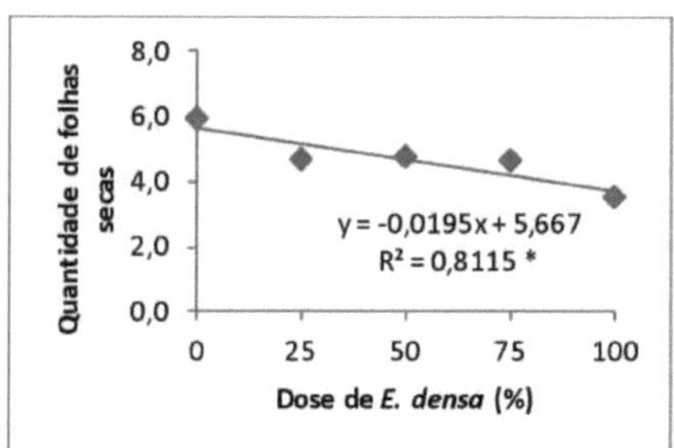

Figure 8. Number of dry leaves of ornamental mini sunflower as a function of doses of the macrophyte *Egeria densa*

Source: author herself.

Another aspect assessed in this study was the number of green leaves, which was higher at an estimated 39.8 per cent of the substrate composed of E. densa (Figure 9). These are important for photosynthesis and influence the development and growth of the plant, as well as being relevant for commercialisation. The number of green leaves was statistically similar between the seaweed treatments, but those that received osmocote fertilisation had a greater number of green leaves (Table 6).

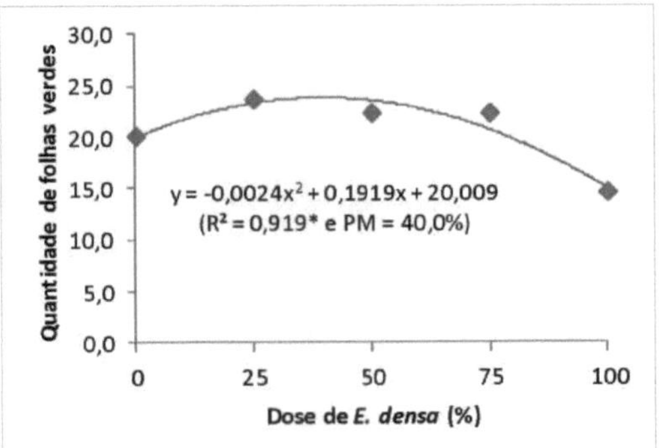

Figure 9- Number of green leaves of ornamental mini sunflower as a function of doses of the macrophyte Egeria densa.

Source: author herself.

The data presented (Table 7) showed that there was a significant difference between the treatments that received mineral fertiliser and those without (with macrophyte only), demonstrating that the fertilised treatments had a greater amount of dry matter. This is due to the faster availability of nutrients for ornamental sunflowers, which have a short cycle, so with greater nutritional availability in the period when the plant needs it most, it is able to develop with greater growth and accumulation of dry matter.

Table 7. Amounts of dry matter (DM) of the chapter, floral stem + leaves, roots and total of ornamental mini sunflower as a function of doses of the macrophyte Egeria densa and mineral fertiliser.

	Chapter DM (g)	**MS flower stem + leaves (g)**	**Root DM (g)**	**Total plant DM (g)**
Doses of E. densa **(% of volume) substrate)**				
0	11,443	16,347	39,303	67,093

25	9,342	13,233	30,481	53,056
50	11,885	14,938	33,259	60,082
75	8,525	11,916	28,141	48,583
100	3,262	2,768	5,628	11,658
Fertilisation				
Without	6,364 b	8,491 b	22,040 b	36,894 b
Conventional	12,134 a	16,444 a	36,160 a	64,738 a
Osmocote	12,658 a	17,306 a	38,734 a	68,698 a
MSD (5%)	3,309	3,750	9,223	10,57
Overall average	9,860	13,346	30,920	54,126
CV (%)	37,890	31,730	33,680	22,05

Means followed by the same letter in the column do not differ according to Tukey's test at 5% probability.

The dry matter of the ornamental sunflower varied depending on the fertiliser (Figure 10). It was found that the dry matter of the chapter and leaves + floral stem were slightly higher in the treatments with 50% E. densa, but in the control there was more dry matter in the roots and total. This is related to the size of the plant, as discussed above. The treatment with 100% E. densa had a lower amount of dry matter, probably due to the smaller volume of substrate at the end of the cycle and the phytotoxicity caused by excess manganese.

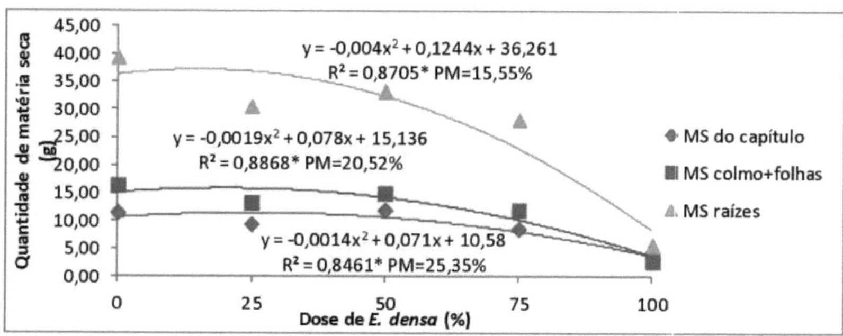

Figure 10- Amount of dry matter in the chapter, leaves+flower stalk and roots of ornamental sunflower as a function of doses of the macrophyte Egeria densa.
Source: author herself.

4.2. NUTRITIONAL ANALYSES

Nutrient analyses were carried out on the leaves + floral stalk and chapter of the ornamental mini sunflower. Firstly, the nutrient content of the leaves and flower stalk will be analysed (Table 8). With regard to fertilisation, it can be seen that only the phosphorus and calcium contents showed a significant effect, with the P content being higher with the application of fertilisers, while the Ca content was higher in the treatments that did not receive any type of fertiliser.

Table 8. Macronutrient content in the leaves + flower stalk of ornamental mini sunflower as a function of doses of the macrophyte Egeria densa and mineral fertilisation.

	N(g/kg)	P (g/kg)	K (g/kg)	Ca (g/kg)	Mg (g/kg)	S (g/kg)
Doses of E. *densa* **(% of substrate volume)**						
0	7,917	3,483	17,500	17,850	7,050	1,883
25	8,500	3,917	21,150	17,050	6,750	1,533
50	11,833	4,317	25,117	15,033	6,483	1,867
75	10,400	3,250	20,475	17,525	7,800	2,000
100	20,650	1,350	9,350	17,800	9,200	2,750
Fertilisation						
Without	10,410 a	2,690 b	18,600 a	19,560 a	7,780 a	1,760 a
Conventional	9,313 a	3,912 ab	18,275 a	15,212 b	6,800 a	1,862 a
Osmocote	12,300 a	4,633 a	25,167 a	14,667 b	6,517 a	2,117 a
MSD (5%)	4,952	1,830	11,577	4,016	1,998	0,491
Overall average	10,517	3,583	20,133	16,886	7,136	1,883
CV (%)	34,530	37,440	42,130	17,430	20,520	19,120

Means followed by the same letter in the column do not differ according to Tukey's test at 5% probability.

The highest nitrogen levels were found in the treatments in which the volume of Egeria densa was greater (Figure 11). This nutrient is responsible for the formation of chlorophyll to carry out photosynthesis and is the nutrient most needed by the plant. It was noted that the greater amount of macrophyte caused a greater amount of this nutrient to be absorbed, which directly affected the amount of senescent leaves and green leaves in the ornamental mini sunflower.

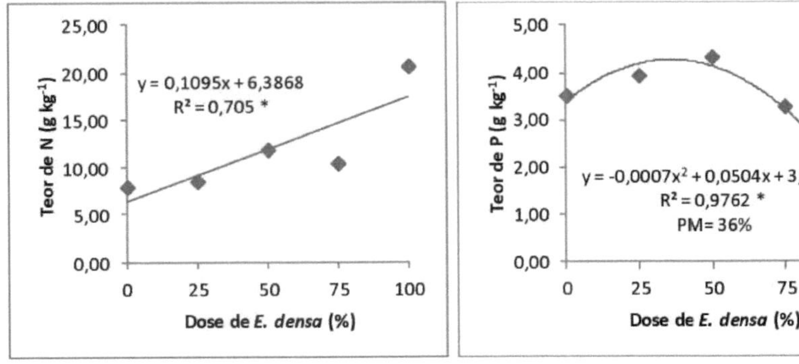

Figure 11-Nitrogen content in leaves + flower stalk as a function of doses of the macrophyte *Egeria densa.*

Figure 12-Phosphorus levels in leaves+flower stalk as a function of doses of the macrophyte *Egeria densa.*

Source: author

The maximum phosphorus content of the mini ornamental sunflower was obtained with the estimated application of 36% macrophyte in the substrate (Figure 12). This nutrient stimulates the growth and formation of the root system at the beginning of the plant's development; this also explains why this treatment had a higher amount of root dry matter compared to the others, as well as a higher plant height. The application of fertilisers showed a statistical difference due to the amount of this nutrient present in the fertilisers, making it more available for the plant to absorb.

Potassium is responsible for enzymatic reactions, as well as being closely linked to photosynthesis and carbohydrate production in the plant. This relationship is apparently due to the influence of K in controlling leaf size, i.e. the smaller the leaf, the lower the carbohydrate production in the plant. With the data obtained (Figure 13), it was observed that the optimum dose estimated was 41.8% of the subtract, in order to obtain the highest carbohydrate content.

of potassium, resulting in a greater amount of dry matter in the flower stalks and leaves.

The highest Ca levels were found in the treatments with the greatest amount of *Egeria densa* (Figure 14). This is due to the fact that this species contains a large amount of Ca in its structure. Calcium promotes and improves root growth, increases microbial activity, and plants with high Ca levels resist Al, Cu and Mn toxicity better.

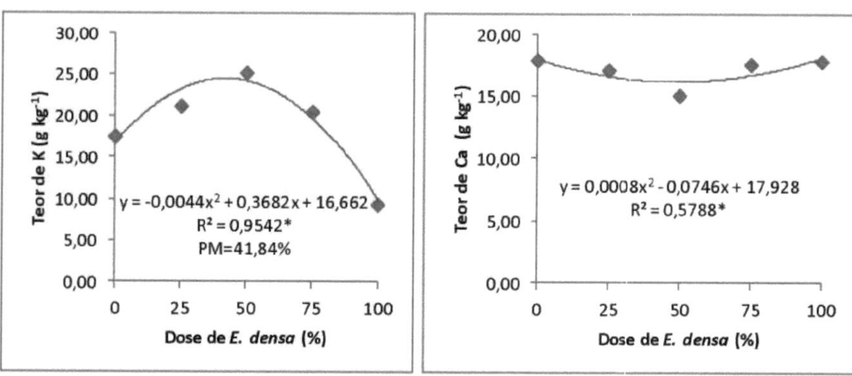

Figure 13- Potassium content in leaves + flower stems as a function of doses of *Egeria densa*

Figure 14- Calcium content in leaves+flower stalk as a function of *doses of Egeria densa*

Source: author herself.

Magnesium is an important nutrient in the constitution of chlorophyll, and with the increase in the amount of seaweed there was also an increase in the Mg content in the leaves and flower stalk (Figure 15), which may have reflected in greater growth of the ornamental mini sunflower.

The sulphur (S) content increased linearly as the doses of macrophyte as substrate increased

(Figure 16). When present in the soil, this nutrient is easily leached and, as it is contained in the composition of *Egeria densa*, considering decomposition/mineralisation, it ends up becoming available more gradually for plant absorption.

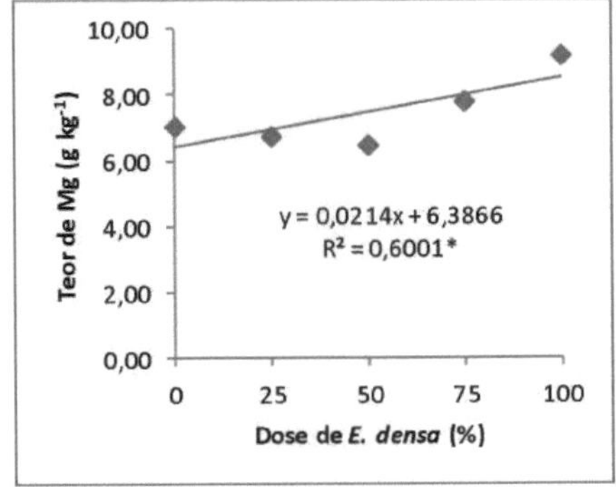

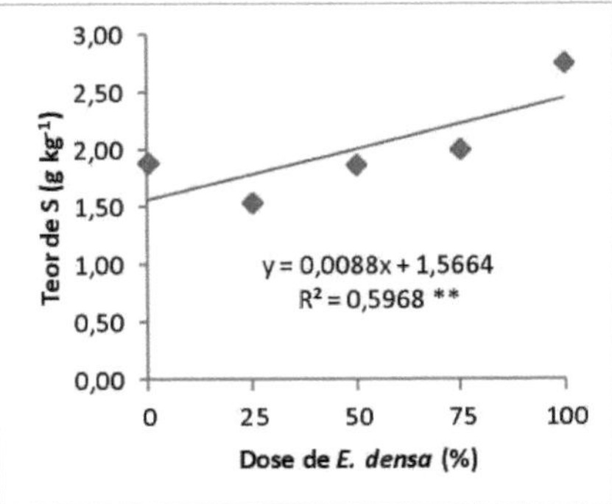

Figure 15- Magnesium content in leaves + flower stems as a function of doses of the macrophyte *Egeria densa.*

Figure 16- Sulphur content in leaves+flower stalk as a function of doses of the macrophyte *Egeria densa.*

Source: author herself.

The results of the micronutrient levels found in the flower stalks and leaves of the ornamental mini sunflower are discussed below (Table 9).

Table 9. Micronutrient content in the leaves + flower stalk of ornamental mini sunflower as a function of doses of the macrophyte Egeria densa and mineral fertilisation.

	B (mg/kg)	Cu (mg/kg)	Fe (mg/kg)	Mn (mg/kg)	Zn (mg/kg)
Doses of E. densa **(% of substrate volume)**					
0	55,667	8,167	135,667	175,000	57,167
25	38,833	5,333	75,500	165,667	43,000
50	42,500	5,833	248,167	411,500	46,500
75	49,750	4,250	226,750	881,500	37,500
100	44,000	4,000	105,500	1105,000	34,000
Fertilisation					
Without	53,500 a	5,900 a	135,500 a	538,600 a	44,500 a
Conventional	41,375 b	5,375 a	173,625 a	424,000 a	40,625 a
Osmocote	40,500 b	6,500 a	125,000 a	245,167 b	54,667 a
MSD (5%)	9,671	2,633	172,735	152,737	29,353
Overall average	46,208	5,875	145,580	427,042	45,750
CV (%)	15,350	32,850	37,930	26,220	24,410

Means followed by the same letter in the column do not differ according to Tukey's test at 5% probability.

As the doses of *Egeria densa* increased, there was no significant adjustment to the B content, but there were lower levels compared to the control (Figure 17). Boron plays a role in cell division, carbohydrate and water metabolism and protein synthesis. A deficiency of this micronutrient impairs growth and can also cause flower buds to fall off. All the treatments, except for the 25 per cent treatment, had sufficient levels for ornamental sunflower nutrition. It was noted that those that received no fertiliser had higher boron levels than those that received fertiliser.

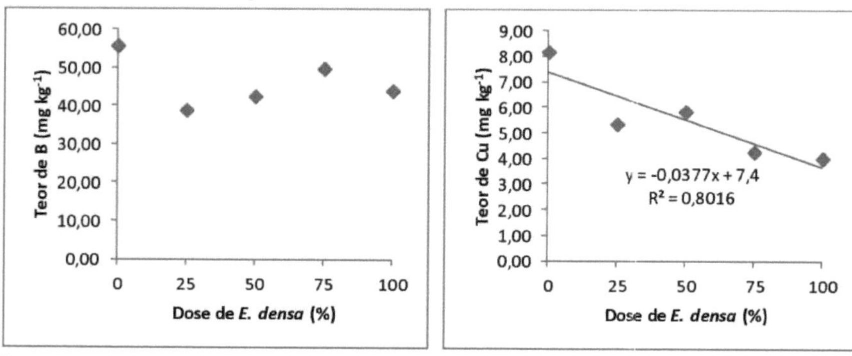

Figure 17: Boron content in the leaves + flower stalk as a function of doses of the macrophyte *Egeria densa.*

Figure 18. Copper content in leaves+flower stalk as a function of doses of the macrophyte *Egeria densa.*

Source: author herself.

Copper is important in the formation of chlorophyll and is needed in small quantities. Its deficiency can cause leaf senescence. It was noted that the content of this micronutrient was inversely proportional to the amount of seaweed used as a substrate (Figure 18), which can be explained by the greater supply and absorption of Mn as a result of the increase in doses of *Egeria densa*, since these compete for absorption sites. However, even with the lower Cu levels, there was no interference in the amount of dry leaves, showing that the amount of this micronutrient was sufficient to meet the ornamental plant's needs.

The Fe content was also not influenced by the increase in macrophyte doses. However, the content was lower when 25% of the substrate contained *Egeria densa* (Table 9). Iron is important for the production of chlorophyll and for the respiration process. The highest levels of this micronutrient were found in the 50% treatment with *Egeria densa.* There was no difference between the use or non-use of fertilisers (Table 9).

It was observed that the Mn content in the plant was high (Figure 19). Mn acts in the synthesis of chlorophyll and participates in energy metabolism. Its deficiency leads to a decrease in photosynthesis and productivity. The Mn content increased as the volume of *Egeria densa* as a substrate increased. This was due to the high amount of Mn present in the macrophyte and, as a result, the plant absorbed more of it.

Zinc is essential for the synthesis of proteins and auxin, the development of flower parts, grain and seed production and early plant maturation. The Zn content decreased linearly with the increase in the macrophyte dose (Figure 20), which again can be explained by the antagonistic effect of the high supply of Mn from *Egeria densa.*

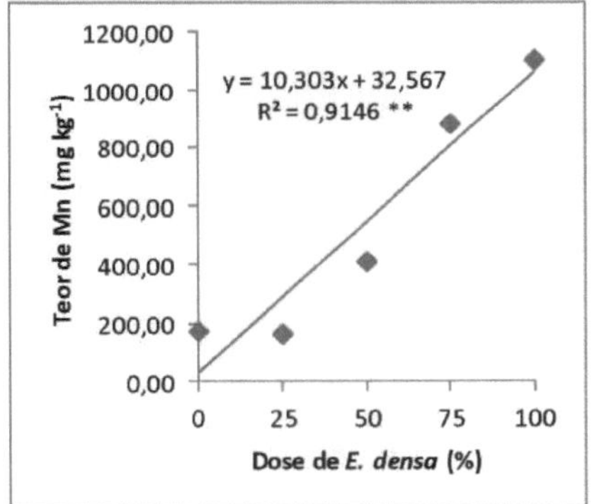

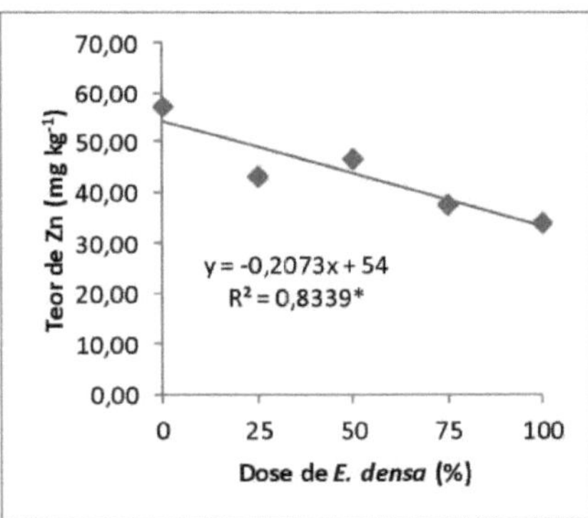

Figure 19- Manganese content in leaves + flower stems as a function of doses of the macrophyte *Egeria densa.*

Figure 20-Zinc levels in leaves+flower stalk as a function of doses of the macrophyte *Egeria densa.*

Source: author herself.

The results of the nutritional analyses of the mini ornamental sunflower chapter are shown in Tables 10 and 11. Regarding the application of fertilisers or not, there was no significant difference for the macronutrients in the inflorescence of the mini ornamental sunflower, demonstrating the potential of this macrophyte as an organic fertiliser.

In the ornamental mini sunflower plant, the highest potassium levels were observed in the leaves+flower stalk and chapter. The decreasing order of nutrient content in the leaves+flower stalk was K, Ca, N, Mg P, S, Mn, Fe, B, Zn and Cu, while in the chapter the descending order of nutrient contents was K, N, Ca, P, Mg S, Mn, Fe, B, Zn and Cu (Tables 8, 9, 10 and 11).

Table 10. Macronutrient contents in the chapters of ornamental mini sunflower as a function of doses of the macrophyte Egeria densa and mineral fertiliser.

	N (g/kg)	P (g/kg)	K (g/kg)	Ca (g/kg)	Mg (g/kg)	S (g/kg)
Doses of E. densa **(% of substrate volume)**						
0	11,767	4,083	20,471	7,483	3,483	1,883
25	13,517	4,167	20,233	7,733	3,167	1,483
50	14,100	4,267	19,717	6,317	3,283	1,517
75	15,025	4,650	21,200	6,350	3,975	1,575
100	22,900	2,950	17,700	4,800	6,250	2,100
Fertilisation						
Without	14,340 a	3,980 a	19,890 a	7,580 a	4,020 a	1,570 a
Conventional	13,750 a	4,338 a	20,412 a	6,100 a	3,536 a	1,700 a
Osmocote	14,800 a	4,183 a	20,033 a	6,600 a	3,250 a	1,750 a
MSD (5%)	4,307	0,869	2,442	1,861	1,089	0,398
Overall average	14,258	4,150	20,100	6,842	3,667	1,658
CV (%)	22,150	15,350	8,910	19,940	21,780	17,600

Means followed by the same letter in the column do not differ according to Tukey's test at 5% probability.

Potassium (K) (Figure 21) is the nutrient with the highest content compared to the others, as it is responsible for the development of the inflorescence. So if the content is low, the plant could lose its commercial value because its flower is not attractive to the consumer.

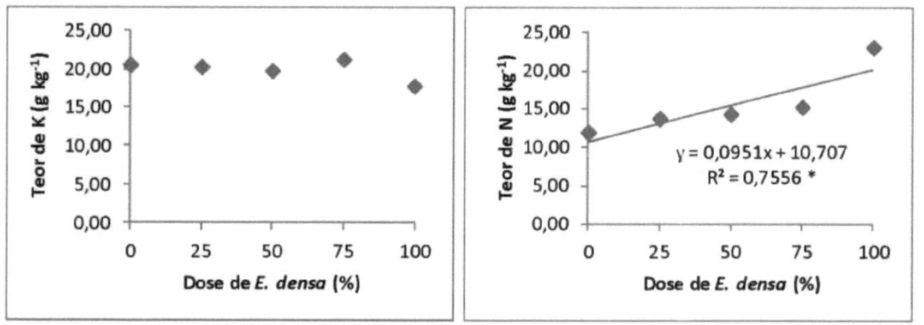

Figure 21-Potassium content in the chapter as a function of doses of the macrophyte *Egeria densa.*

Figure 22- Nitrogen content in the chapter as a function of doses of the macrophyte *Egeria densa.*

Source: author herself.

The second most absorbed nutrient in the chapter was nitrogen, the effects of which have already been discussed. In this case, it was noted that the N content increased as the dose of E. *densa* increased. Therefore, the more macrophyte supplied, the greater the plant's absorption (Figure 22). The next most absorbed nutrient is calcium (Figure 23), which is absorbed for the purpose of producing future seeds by providing a rigid layer of protection.

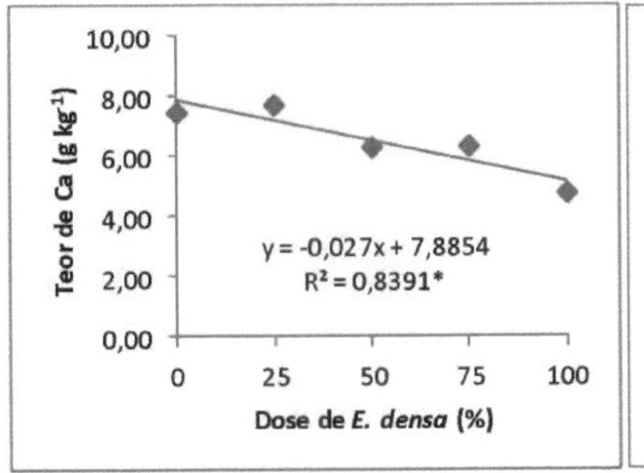

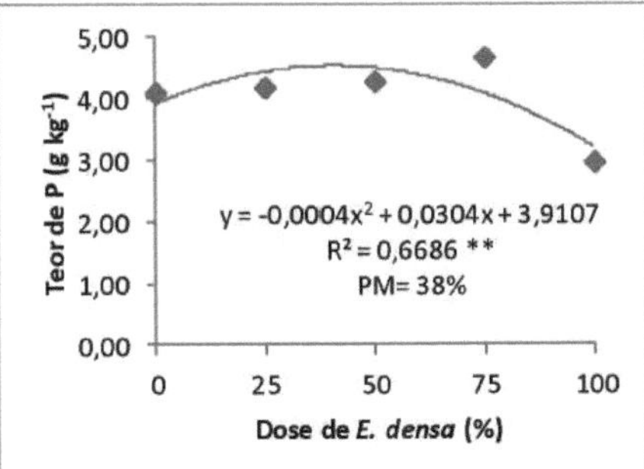

Figure 23- Calcium content in the chapter as a function of doses of the macrophyte *Egeria densa.*

Figure 24- Phosphorus content in the chapter as a function of doses of the macrophyte *Egeria densa.*

Source: author herself.

As for the other macronutrients P, Mg and S (Figures 24, 25 and 26), the amount absorbed was similar. However, for the 100 per cent treatment with *Egeria densa*, phosphorus decreased absorption, while the opposite was true for Mg and S.

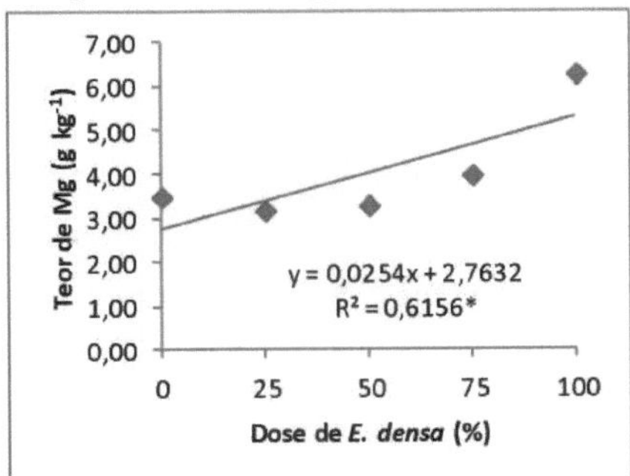

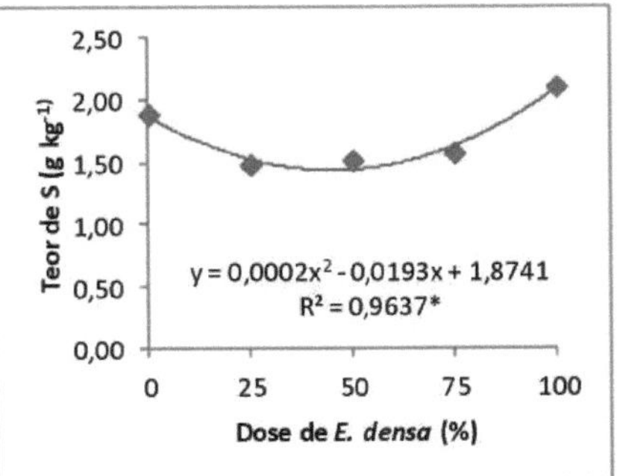

Figure 25- Magnesium content in the chapter as a function of doses of the macrophyte *Egeria densa.*

Figure 26- Sulphur content in the chapter as a function of doses of the macrophyte *Egeria densa.*

Source: author herself.

For the micronutrients in the chapter (Table 11), only the boron and manganese levels differed significantly in relation to fertiliser, with the highest B content being obtained without mineral fertiliser. The Mn content was lower when the osmocote fertiliser was applied.

Table 11. Micronutrient content in the chapters of ornamental mini sunflower as a function of doses of the macrophyte Egeria densa and mineral fertilisation.

	B (mg/kg)	Cu (mg/kg)	Fe (mg/kg)	Mn (mg/kg)	Zn (mg/kg)

Doses of E. densa **(% of substrate volume)**					
0	27,500	9,667	21,333	53,500	25,833
25	23,167	7,333	25,833	50,333	21,000
50	25,000	6,667	34,167	119,333	25,500
75	21,500	5,500	24,500	204,750	25,250
100	24,000	5,500	33,000	215,500	34,000
Fertilisation					
Without	26,700 a	6,800 a	33,900 a	112,500 a	22,300 a
Conventional	21,750 b	6,375 a	22,265 a	124,625 a	26,375 a
Osmocote	24,500 ab	9,333 a	22,000 a	77,833 b	28,167 a
MSD (5%)	3,050	4,086	19,645	30,792	12,178
Overall average	24,500	7,292	27,167	107,875	21,125
CV (%)	9,130	41,090	53,020	20,930	35,540

Means followed by the same letter in the column do not differ according to Tukey's test at 5% probability.

The ornamental mini sunflower is a plant that is sensitive to boron deficiency, causing poor flower formation. We can see that in this case (Figure 27) the treatments without fertiliser provided greater availability of this nutrient to be absorbed, which means that the macrophyte was sufficient for the development of the plant under study.

The copper absorbed by the sunflower flower (Figure 28) was inversely proportional to the increase in the dose of Egeria densa used as a substrate, while for the micronutrients Fe, Mn and Zn the levels increased as the amount of macrophyte supplied increased (Figures 29, 30 and 31).

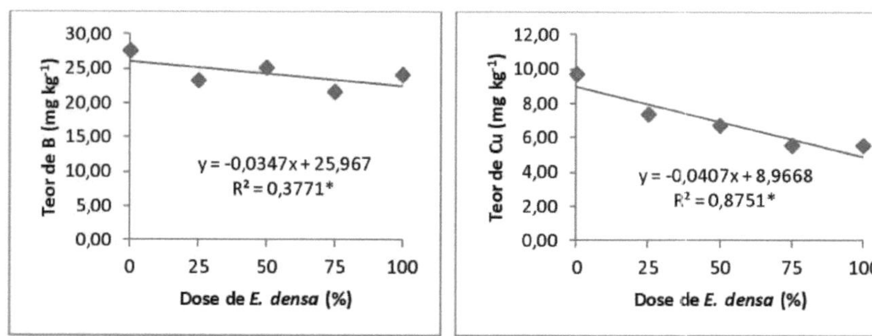

Figure 27-Boron **levels** in the chapter as a function of doses of the macrophyte *Egeria densa.*

Figure 28- Copper content in the chapter as a function of doses of the macrophyte *Egeria densa*

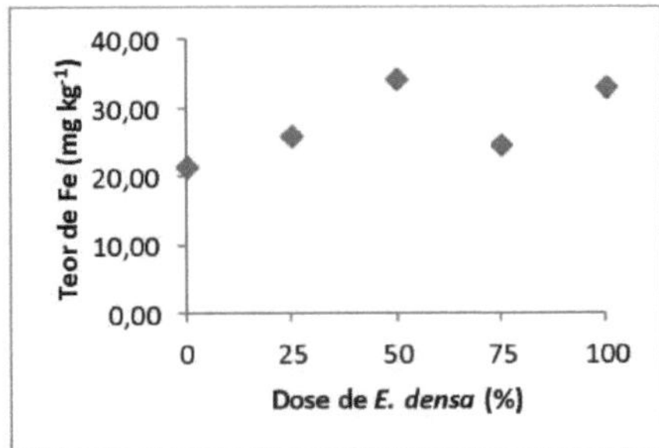

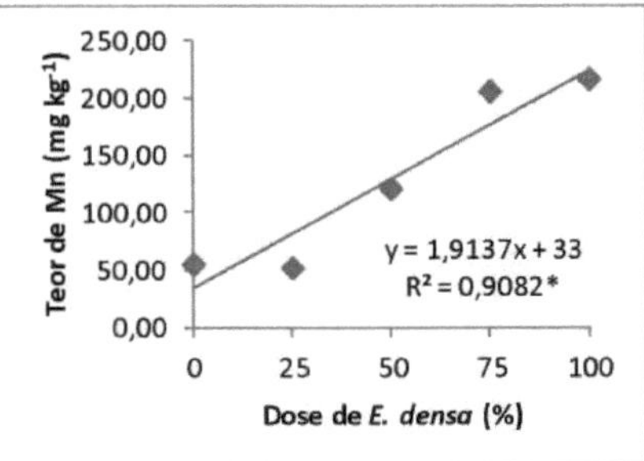

Figure 29- Iron content in the chapter as a function of doses of the macrophyte *Egeria densa.*

Figure 30- Manganese content in the chapter as a function of doses of the macrophyte *Egeria densa.*

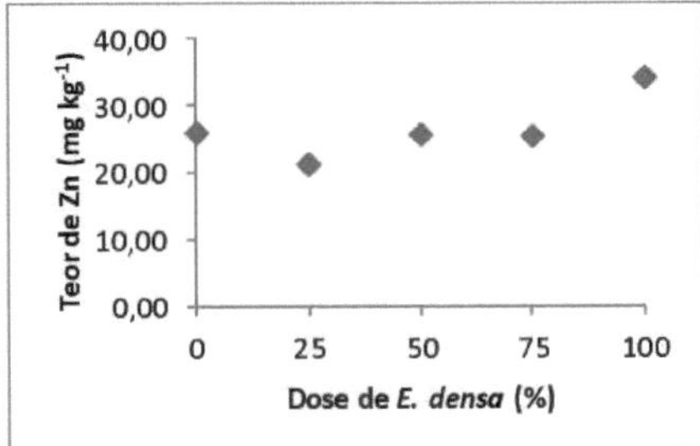

Figure 31 - Zinc **levels** in the chapter as a function of doses of the macrophyte Egeria densa

Source: author herself.

All the parameters analysed and discussed in this paper are important for obtaining an ornamental plant of good quality and standard, which comes from an adequate supply of nutrients.

All the quantitative and qualitative analyses showed that Egeria densa as a substrate for ornamental sunflower production had a positive influence on its appearance. Therefore, methods and technologies should be sought that are aimed at sustainability, as well as production and reducing production costs. To this end, alternative substrates should be studied, such as this macrophyte, which was promising, but which should not be used entirely without mineral fertilisation.

CHAPTER 5

CONCLUSIONS

Increasing the doses of Egeria densa macrophyte in the substrate had a positive effect on the levels of N, P, K (leaves+flower stalk), Mg, S, Mn, ICF, plant height, flower stem and flower disc diameters, vegetative and root dry matter of ornamental sunflower, regardless of mineral fertilisation. However, the highest doses of this macrophyte reduced Cu and Zn levels (leaves+flower stem) and root dry matter.

The use of the macrophyte Egeria densa as a nutrient substrate in a proportion of between 25 and 40% was feasible for the production of ornamental sunflowers with standards suitable for marketing both in pots and for cutting, provided that it was associated with conventional mineral fertilisation or with osmocote.

CHAPTER 6

REFERENCES

BECALETO, B. D.; TEIXEIRA FILHO, M. C. M.; FREITAS, L. A; GARCIA, C. M. P.; BUZETTI, S. Growth and dry matter production of maize plants as a function of doses of seaweed (Egeria densa) with and without nitrogen application. In: Unesp Scientific Initiation Congress, 25, 2014. Ilha Solteira. **Proceedings...** Ilha Solteira: UNESP, 2014. Available at: < http://prope.unesp.br/cic_isbn/busca.php>. Accessed on: 13 Dec. 2014.

BRAGA, C. L. **Doses of nitrogen in the development of potted ornamental sunflower** (Helianthus annuus **L.)**. Botucatu, São Paulo. Master's dissertation. Faculty of Agronomic Sciences, Unesp, 91p. 2009.

CARVALHO, S. A. Citrus propagation. In: Citriculture: technological innovations. **Informe Agropecuário**, Belo Horizonte, v.22, n.209, p.21-25, 2001.

CORREA, M. R.; VELINI, E. D.; ARRUDA, D. P.Chemical and bromatological composition of Egeria densa, Egeria najas and Ceratophyllum demersum. **Planta Daninha.** Viçosa, v.21, n.spe, p.7-13, 2003.

CURTI, G. L. **Characterisation of ornamental sunflower cultivars sown at different times in western Santa Catarina**. Pato Branco, Paraná. Master's dissertation. Technological University of Paraná, 76p. 2010.

CURTI, G. L.; MARTIN, T. N.; FERRONATO, M. L.; BENIN, G. Ornamental sunflower: characterisation, post-harvest and senescence scale. **Revista de Ciências Agrárias**, Lisboa, v.35, n.1, p-240-250, 2012.

DOORENBOS, J.; KASSAM, A. H. **Efectos del agua solbre el rendimiento de los cultivos** . Rome, 1979. 212p. (FAO. Irrigation and Drainage, 33).

EVANGELISTA, A. R.; LIMA, J. A. **Sunflower silage: cultivation and ensiling**. Available at <editora.ufla.br/upload/boletim/extensao-tmp/boletim-extensao-087.pdf>. Accessed on: 18 October

2015.

FERREIRA, R. N. D.; BELO, M. **Cadeia produtiva da floricultura no Estado do Rio de Janeiro**. Nova Friburgo, RJ: SEAPEC/EMATER-RIO - Secretaria de Estado de Agricultura e Pecuária/ Empresa de Assistência Técnica e Extensão Rural no Estado do Rio de Janeiro, p.167. 2015.

FRANÇA, C. A. M.; MAIA, M. B. R. Panorama of the flower and ornamental plant agribusiness in Brazil. In: Congress of the Brazilian Society of Economics, Administration and Rural Sociology; 46, 2008. Rio Branco. **Proceedings...** Rio Branco: SOBER, 2008.

HIGAKI, T.; IMAMURA, J. S.; PAULL, R. E. N, P and K rates and leaf tissue standarts for optimum Anthurium andraenaum flower production. **Hortscience**, University of Hawaii, v.27, n.8, p.909-912, 1992.

IBRAFLOR, Brazilian Institute of Floriculture. Available at: <http://www.ibraflor.com/publicacoes/vw.php?cod=246>. Accessed on 10 October 2015.

IBRAFLOR, Brazilian Institute of Floriculture. **Mapping and Quantification of the Brazilian Flower and Ornamental Plant Chain**. São Paulo: OCESP, p.132. 2015.

JUNQUEIRA, A. H.; PEETZ, M. S. Domestic market for Brazilian floriculture products: characteristics, trends, recent socioeconomic importance. **Revista Brasileira de Horticultura Ornamental**, Campinas, v.14, n.1, p.37-52, 2008.

LEMAIRE, P. Physical, chemical and biological properties of growing medium. **Acta Horticulturae**, v.396, n.1, p.273-284, 1995.

LENTZ, D.; POHL, M. E. D.; POPE, K. O.; WYATT, A. R. Prehistoric sunflower (Helianthus annuus L.) domestication in Mexico. **Economic Botany**, 2001. 370-376 p.

MALAVOLTA, E.; VITTI, G. C.; OLIVEIRA, S. A. **Avaliação do Estado Nutricional das Plantas:** princípios e aplicações (2ª . edição), Potafos, Piracicaba, SP. 1997, 319p.

MÓDENES, A. N.; ESPINOZA, F. R.; ALFLEN, V. L.; COLOMBO, A.; BORBA, C. E. Use of the macrophyte Egeria densa in the biosorption of the reactive dye 5G. **Engevista**, v.13, n.3., p.160-166, 2011.

NASCIMENTO, P. R. F.; SAMPAIO, E. V. S. B., PEREIRA, S. M. B., MOURA JUNIOR, A. M 2001. Evaluation of the growth and biomass of Egeria densa Planchon (Hydrocaritaceae), in the hydroelectric reservoirs of Paulo Afonso - Bahia. In: VIII Brazilian Congress of Limnology: Biodiversity and Water Resources, João Pessoa - PB. **Proceedings...** João Pessoa, p.47.

NEVES, M. B. **Development of ornamental sunflower plants** (Helianthus annuus **L.) in pots, in two substrates, with nutrient solution and in soil**. Ilha Solteira. Master's dissertation. Faculty of Engineering of Ilha Solteira, Paulista State University. p.63, 2003.

NEVES, M. B.; ANDRÉO, Y. S., WATANABE, A. A.; FAZIO, J. L.; BOARO, C. S. F. Use of daminozide in the production of ornamental sunflower grown in pots. **Revista Eletrônica de Agronomia**, Garça, v.16, n.2, p.31-37, 2009.

OLIVEIRA, M. F.; CASTIGLIONI, V. B. R. **Coloured sunflower for Brazil**. Londrina. EMBRAPA- CNPSO, 2003.

OLIVEIRA, A. B; HERNANDEZ, F. F.; JUNIOR ASSIS, R. N. Green coconut dust, an alternative substrate for the production of aubergine seedlings. **Revista Ciência Agronômica**, Fortaleza, v.39, n.1, p.39-44, 2008

PELEGRINI, B. **Sunflower: a solar plant that conquered the world from the Americas**. São Paulo: Ícone. 1985, 117 p.

PIERINI, S. A.; THOMAZ, S. M. Adaptations of submerged plants to the absorption of inorganic carbon. **Acta Botanica Brasílica**, São Paulo, v. 18, n. 3, p.629-641, 2004.

PRINCIPE, C. R.; KURATANI, H.; MELONI, M. L. B. Impact of the influx of eelgrass on the operation and maintenance of the Eng. Souza Dias (Jupiá) hydroelectric power station - CESP. In: Congress

Brasileiro de Ciência das Plantas Daninhas; 21, 1997. Caxambu. Workshop on aquatic plants. **Proceedings...** Caxambu: SBCPD, p. 5-8, 1997.

RIBEIRO, M. C. C.; GURGEL JUNIOR, C. A.; MENDES, V. H. C.; BENEDITO, C. P.; OLIVEIRA, G. L.; NUNES, T. A.; FIGUEREDO, M. L. Use of the growth retardant paclobutrazol in sunflower (Helianthus annuus). **Revista Brasileira de Biociências,** Porto Alegre, v. 5, supl. 2, p.1104-1106. 2007.

RODRIGUES, R. B.; DETTKE, G. A.; MONTANHER, D. R. Anatomy of species of the genera Egeria Planch. and Hydrilla Rich. (Hydrocharitaceae). **Revista Brasileira de Biociências**, Porto Alegre, v.5, supl.1, p.360-362, 2007.

SALUNKHE, D. K.; DESAI, B. B. Sunflower. In: SALUNKHE, D. K.; DESAI, B. B. **Postharvest biotechnology of oilseeds**. Boca Raton: CRC Press, 1986. p.57-92.

SAMPAIO, E .V. S. B.; OLIVEIRA, N. M. B; NASCIMENTO, P. R. F. Efficiency of organic fertilisation with cattle manure and Egeria densa. **Revista Brasileira de Ciência do Solo**, Viçosa, v.31, n.5, p.995-1002, 2007.

SAMPAIO, E. V. S. B.; OLIVEIRA, N. M. B. Use of the aquatic macrophyte Egeria densa as an organic fertiliser. **Planta Daninha,** Viçosa, v.23, n.2, p.169-174, 2005.

SCOTTS UK PGB. **Osmocote®: The working principle of osmocote**. United Kingdom, 2002. Available at: <http://www.osmocote.co.za/about.htm>. Accessed on 10 January 2015.

SFREDO, G. J.; CAMPOS, R. J.; SARRUGE, J. R. **Sunflower: mineral nutrition and fertilisation**. EMBRAPA-CNPS, 1984. 36 p.

SILVA, G. D.; MANO, A. R. O. ; MAIA, M. R. P.; SILVA, W. S.; SOUSA, M. G. F. Efficiency of organic fertilisation with macrophyte for the production of aroeira seedlings *(Myracrodruon urundeuva* Fr. Allem). In: 64th NATIONAL CONGRESS OF BOTANICS. **Proceedings...** Belo Horizonte. 2013.

SILVA, J. V. H.; BORGES, A. K. P.; MORAIS, P. B.; PICANÇO, A. P. **Utilisation of aquatic macrophytes removed from the reservoir of the Luis Eduardo Magalhães HPP, Tocantins, in the composting process and evaluation of the usefulness of the compost**. In: 3rd Ibero-American Waste Engineering Symposium and 2nd Northeast Region Seminar on Solid Waste. 2009. 8p.

SILVA, L. C. **Characterisation of the flower and ornamental plant wholesale sector in Brazil**. Lavras, Minas Gerais. Master's dissertation. Federal University of Lavras, 2012. 135p.

SIMÃO, M .L. **The sunflower** (Hellianthus annuus) **for cut flowers (in line)**. 2004. Centro de Experimentação de Horticultura da Gafanha, DRAPC - Direção Regional de Agricultura e PescasdoCentro . Disponível:<http:// www.drapc.minagricultura.pt/base/documentos/girassol_flor_corte.htm>. Accessed 16 October 2015.

UNGARO, M. R. G. **Instructions for sunflower cultivation**. Campinas: IAC, 1986, 26p. (Technical Bulletin 105).

VALERI, S. V.; CORRADINI, L. Fertilisation in nurseries for the production of Eucalyptus and Pinus seedlings. In: GONÇALVES, J. L. M.; BENEDETTI, V. **Nutrição e fertilização florestal**. Piracicaba: IPEF, 2000. p.168-190.

WETZEL, R. G. **Limnologia**. 2ª Ed., Saunders College Publishing, Lisbon. 1993. 919p.

ZAMUNÉR FILHO, A. N. **Doses of slow-release fertiliser for the production of rubber tree rootstocks**. Master's dissertation. Federal University of Lavras, Lavras - MG. 2009. 45p.

Printed by Books on Demand GmbH, Norderstedt / Germany